Matters of Opinion

RECYCLING

By
CARLA MOONEY

NORWOOD HOUSE PRESS
CHICAGO, ILLINOIS

Norwood House Press
P.O. Box 316598
Chicago, Illinois 60631

For information regarding Norwood House Press, please visit our website at:
www.norwoodhousepress.com or call 866-565-2900.

LIBRARY OF CONGRESS CATALOGING-IN-PUBLICATION DATA

Mooney, Carla, 1970-
  Recycling / by Carla Mooney.
    pages cm
  Includes bibliographical references and index.
  Summary: "Explores the pros and cons of several issues related to
recycling, including whether it helps the environment, and is it
economically efficient. Aligns with Common Core Language Arts Anchor
Standards for Reading Informational Text and Speaking and Listening. Text
contains critical thinking components in regards to social issues and
history. Includes bibliography, glossary, index, and relevant websites"--
Provided by publisher.
  ISBN 978-1-59953-603-3 (library edition : alk. paper) -- ISBN
978-1-60357-596-6 (ebook) 1. Recycling (Waste, etc.)--Juvenile literature.
I. Title.
  TD794.5.M66 2014
  363.72'82--dc23
                                        2014005102

252N—072014
Manufactured in the United States of America in Stevens Point, Wisconsin.

Contents

Note: Words that are **bolded** in the text are defined in the glossary.

Timeline

1965 ▸ The U.S. government enacts the Solid Waste Disposal Act.

1970 ▸ The first Earth Day brings national attention to the problem of increasing waste and the importance of recycling. The U.S. Environmental Protection Agency (EPA) is created.

1972 ▸ The federal Clean Water Act is enacted to restore and maintain the integrity of the nation's waters.

1976 ▸ The federal Resource Conservation and Recovery Act is enacted to close open dumps and create standards for landfills, incinerators, and the disposal of hazardous waste. The Toxic Substances Control Act is passed.

1990 ▸ McDonald's restaurants stop using Styrofoam containers. Coca-Cola and PepsiCo announce that they will begin using a recycled PET (plastic) bottle made of about 25 percent recycled plastic.

1996 ▸ The EPA sets a new recycling goal of 35 percent, an increase from the U.S. recycling rate of 25 percent.

2000 ▸ The EPA confirms a link between global warming and waste. The agency says that reducing trash and recycling will cut greenhouse gas emissions.

2005 Discarded electronics such as old computers, broken cell phones, and obsolete television sets become the fastest-growing part of the waste stream. Americans throw out 2 million tons (1.8 million t) of electronics in 2005 and only recycle about 380,000 tons (344,730 t).

2006 Dell offers a free recycling service for its electronic products.

2007 Five states pass laws requiring that unwanted electronics be recycled. San Francisco becomes the first U.S. city to prohibit the distribution of plastic bags by grocery stores.

2009 San Francisco implements the most stringent mandatory composting and recycling law in the country. It requires every residence and business to have three separate color-coded bins for recycling, compost, and trash.

2013 The New York City Council votes to ban plastic-foam food service containers citywide. The containers are hard to recycle and usually are thrown away in landfills.

1 Why Is Recycling an Issue?

The EPA says Americans are throwing out more waste than ever. In 1960 they threw out about 88.1 million tons (80 million t) of waste. That is about 2.68 pounds (1.2kg) per person per day. In 2011 the amount of waste was much higher. It grew to about 250 million tons (227 million t). That is about 4.4 pounds (2kg) per person each day. People throw out waste from lots of things. These include packaging and food. People also throw out grass clippings, furniture, electronics, tires, and appliances. Waste comes from homes and businesses. It also comes from schools and hospitals.

Disposable containers only add to the growing amount of waste. Plastic water bottles are one example of this. In 1960 people did not drink bottled water. The Beverage Marketing Corporation says that by 2012 Americans were drinking a lot of bottled water. According to their data, Americans were drinking about 31 gallons (117L)

Waste from things like water bottles can pile up.

per person. Most of those bottles were thrown away. This adds to plastic waste. Today more and more things are packaged and sold in disposable containers. So this waste will just keep growing.

What Happens to Waste?

Different things happen to something that is thrown out. Most waste is sent to landfills. There it is buried underground. The EPA says Americans sent more than half of their

European countries throw out much less waste, from 2 to 3 pounds (907g to 1,361g) of waste per person per day whereas in the USA people throw out about 4.4 pounds (2kg) per person each day.

waste to landfills in 2011. Other waste is incinerated. **Incineration** is a process that burns waste. In some cases this creates energy that can be used by people. In 2011 about 11.7 percent of waste was incinerated.

Organic waste is things like grass clippings, leaves, or food scraps. These can be composted. In **composting**, organic wastes are combined with other things, such as wood chips. The mixture speeds up the

Because the approval process is difficult, no new incineration plants like this one have been built in the United States since 1997.

process of breaking down the wastes. They become a rich soil known as compost. Composting returns necessary nutrients back into the soil. The nutrients support future plant growth.

Other waste is recycled. **Recycling** is a process that converts waste into new, usable products. It reuses useful materials. At the same time, it reduces the need for raw materials. Things that are commonly recycled include glass, paper, metal, plastic, **textiles**, and electronics. To be recycled, waste is brought to a collection center. There it is sorted, cleaned, and sent to be processed into new materials. The EPA says Americans recycled or composted about 34.7 percent of their waste in 2011.

The average American contributes nearly 4.5 pounds a day – 56 tons a year – of garbage that is carted off to landfills like this one.

What Is the Problem?

So what is the problem with all this waste? Most waste ends up in a landfill. This is not a great solution. Many materials dumped into a landfill need centuries to **decompose**. Landfills also take up a lot of space. In many populated places, landfill space is scarce. Few people want to live next to one. Landfills are also a source of pollution. Trash and chemicals break down and generate pollution. This can drain out of the landfill and harm the surrounding air, water, and soil. A landfill's

Recycling Pride

The Environmental Industry Associations (EIA) did a survey in 2013. It found that more than four in five people feel a sense of pride when they recycle. Sharon H. Kneiss is the president and CEO of the EIA. She says these results show that people want to recycle if given the chance. "Recycling participation rates have increased dramatically during the last few decades in the U.S., and that is an achievement that all Americans should celebrate," she says. "There is positive, pent-up desire to recycle even more in America. But we need more recycling options on our main streets … and other public spaces."

Quoted in National Waste & Recycling Association, "Environmental Industry Associations Survey Finds Most Americans Are Proud to Recycle—When They Can," November 13, 2013.

decomposing waste also creates **greenhouse gas** emissions. These can harm the Earth's atmosphere.

Also, for every water bottle thrown into a landfill, the raw materials and energy used to make that bottle are lost. New materials and energy are needed to make a new bottle. New plastics use limited fossil fuels like oil and natural gas. The same thing is true with paper products. New paper products use wood pulp from trees. Also, making new goods uses energy resources to power equipment.

Recycling Solution

Many people think that recycling is a better way to dispose of waste. By reusing items, recycling saves natural resources. In many cases it takes less energy to make a recycled product than it takes to make a new product. So recycling can decrease the demand for energy to make new goods. Also, recycling means less waste goes to landfills. So fewer landfills are needed. And the landfills can be smaller. They will also make fewer greenhouse gases and pollutants.

Many cities and towns across the country promote recycling. They have curbside recycling programs. In these programs people put recyclable waste in special bins on the curb each week. Special trucks pick up the recycling waste. They take it to recycling centers for

A garbage truck collects the trash people throw out each week. There are many ways to reduce waste. One is recycling.

processing. Some recycling programs are voluntary. Others are **mandatory**.

Concerns About Recycling

Recycling is not a perfect solution for waste. Some people think that it promises more than it delivers. They point out that recycling programs are costly to run. Places that recycle must pay for recycling trucks. They pay to build processing plants and equipment. They also pay to hire staff. Recycling can also be more costly and time-consuming than just throwing out trash.

A paper recycling plant. Some believe the cost of recycling is too high.

Some people point out that recycling is also flawed in other ways. The trucks and equipment use energy to operate. They emit greenhouse gases. These cause pollution and harm the planet. Some people think these costs take away from the benefits of recycling.

And the recycling process is not always simple. Some materials, such as plastics, are hard to recycle. Plastics need more processing than glass or metal. They must be carefully sorted first, sometimes by hand. Different types of plastic cannot be mixed together in recycling. If they are, the whole batch becomes unusable. And many plastics contain dyes or other **additives**. This makes them hard to recycle.

Sorting recyclables is a lot of work.

In 2002 New York City halted its recycling of glass and plastics. Michael Bloomberg was the city's mayor then. He says the recycling program was cut because it cost more than it saved. He told the city council, "The recycling program is not, with the exception of paper, saving the ecology of the world very much. And it is very expensive."[1]

Many people spoke out against New York City's decision to end these programs. Suzanne Shepard of the New York chapter of the Sierra Club was one of these people. She said, "To stop recycling would be to turn the clock backward. Recycling and waste reduction are the cornerstones to reducing this city's waste stream."[2] In 2003 Bloomberg restarted New York City's plastics recycling program. And he restarted the glass program in 2004.

A Look Inside This Book

The recycling debate is complex. Since people will always create waste, the debate will continue. In this book, three issues will be covered: Is recycling good for the environment? Is recycling economically efficient? Should recycling be mandatory? Each chapter ends with a section called **Examine the Opinions**, which highlights one argumentative technique used in the chapter. At the end of the book, students can test their skills at writing their own essay on the book's topic. Finally, notes, glossary, a bibliography, and an index provide additional resources.

2 Is Recycling Good for the Environment?

👍 Yes: Recycling Is Good for the Environment

Recycling is a simple idea. Take something that is being thrown out and make it into something new. Recycling can be as simple as turning an old tire into a tree swing. It can also be complex, like turning large amounts of plastic into new goods. Some people think that recycling is a big part of protecting the planet. Patty Moore is a recycling consultant. She helps businesses, governments, and cities with waste management issues. "It helps us conserve resources and energy, and it's something all of us can do every day to help make a difference,"[3] she says.

A man trims a giant roll of recycled paper.

Saves Natural Resources

The National Recycling Coalition says recycling can help the planet. It lowers the amount of waste that goes to landfills. This reduces the need for new landfill space.

Recycling also saves natural resources. When 1 ton (0.9t) of steel is recycled, it saves 1,400 pounds (635kg) of coal and 120 pounds (54kg) of limestone. This is what is needed to make a new batch of steel. Recycling paper saves trees. It provides 37 percent of the raw materials used to make new paper products. It reduces the need to extract, move, and process raw materials. So it also lowers the pollution made by these things. One example is processing minerals. This sends 1.5 million tons (1.4 million t) of pollution into the air and water each year in the United States. Recycling that lowers demand for the minerals also lowers these emissions.

Recycling can also save energy. And it lowers the country's need for foreign oil. In many cases a recycled product can be made using less energy than a new product. Earth911 is a group that promotes recycling.

Recycling Reduces Overall Emissions

Recycling trucks release greenhouse gases. But some people think the net effect of recycling still lowers overall greenhouse gases. David Allaway is a policy and program analyst. He says, "If you compare the greenhouse gas requirements of all those trucks driving around collecting all those materials and making the fuel for the trucks versus the reductions when those materials are recycled, the savings are about 40 times higher than the collection impacts. So, don't sweat recycling collection. It's trivial."

Quoted in Cassandra Profita, "The Impacts of Bottles vs. Cans: Further Analysis," Oregon Public Broadcasting, July 13, 2011.

It says 20 recycled cans can be made with the energy it takes to make one new can. The EPA says making recycled plastic uses two-thirds the energy needed to make new plastic. Recycled paper uses about 60 percent of the energy needed to make paper from wood pulp. In all of these cases, recycling saves energy.

For all of these reasons, many people think that recycling helps the planet. Bob Martin is commissioner

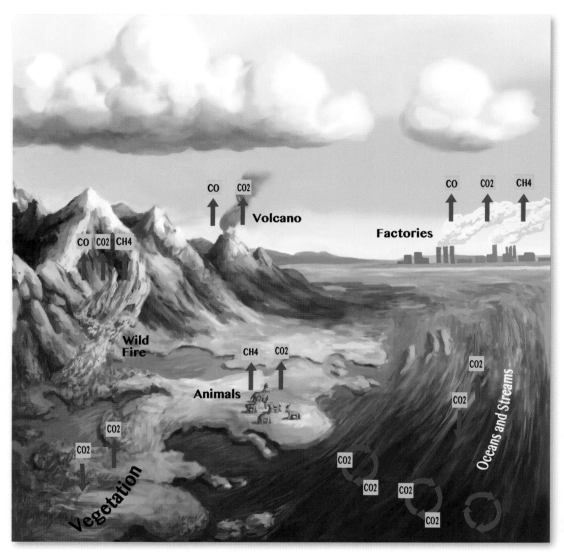

The greenhouse effect happens when a carbon atom combines with two oxygen atoms making carbon dioxide (CO_2). Industrial sites, volcanoes, wildfire, and animals all release carbon gases, which warm the atmosphere.

of New Jersey's Department of Environmental Protection. He says, "Recycling remains one of the best ways for each one of us to be actively engaged in protecting our environment and conserving our natural resources."[4]

Reducing Greenhouse Gases

Fossil fuels are burned to make energy. But burning them puts greenhouse gases in the air. Landfills and incinerators also emit greenhouse gases into the air. These gases have been linked to global warming. Many say that recycling saves energy. It also means fewer landfills and incinerators. So it can reduce greenhouse gas emissions. The South Carolina Department of Health and Environmental Control (DHEC) says a national recycling rate of 35 percent lowers greenhouse gases a lot. It would be the same as taking almost 25 million cars off the road for one year. Susan Collins works for the Container Recycling Institute. She thinks that recycling is a good way to reduce these gases. This is because recycling lowers the emissions that come from mining and manufacturing. "These savings are substantial—the

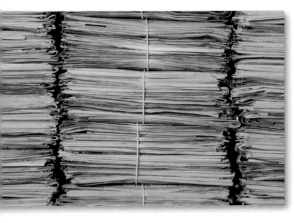

These old newspapers will be recycled into new paper.

carbon footprint of 100% recycled content plastic bottles is about 70–90% lower than that of bottles made from virgin materials,"[5] she says.

The DHEC says there are benefits to its recycling program. Between July 2011 and June 2012, the state recycled 1.2 million tons (1.1 million t) of solid waste. The emissions saved because of this are equal to those from the electricity use of 434,227 homes for one year. It is also equal to saving more than 325 million gallons (1.2 billion L) of gas.

But Not So Fast...

No: Recycling Is Bad for the Environment

Some people say that recycling has many hidden costs. The process of recycling itself uses energy. It also creates pollution. Recycling trucks and centers use energy. They release emissions and greenhouse gases into the atmosphere. Recycling plants emit pollution into the air. These costs can add up. So the overall benefit of recycling may be exaggerated.

And some people disagree that modern landfills are a bad way to dispose of waste. They say landfills are safe for most waste. Drew Thornley is a policy analyst. He focuses on energy and the environment. He says, "Modern landfills are designed to keep air, light, and moisture away from the waste, in essence mummifying the waste to prevent decay and minimize the release of liquids and gases. Small releases (if any) are vented and drained to prevent environmental harm."[6]

Recycling's effect on natural resources may also be exaggerated. One example of this is paper. Most of the pulp from paper comes from

Recycling has hidden costs, including the cost to use separate trucks, such as this one.

sources that are grown for paper. If the demand for paper dropped, the businesses that own these trees would find another way to use them. Todd Myers is director of the Center for the Environment. He says:

> Lower demand for paper would mean less demand for hardwood trees that make paper. Forestland owners, finding no market for their pulpwood trees, would likely convert those forests to some other crop that would produce revenue. The notion that foresters would simply leave valueless trees on the land, continuing to pay property taxes and foregoing revenue from other crops, is not based in reality. Lower demand for hardwood trees is likely to mean fewer hardwood trees, not more.[7]

Find a Better Way

Other critics think that focusing on recycling is not the best idea. They say it takes away from a better way to protect the planet. They say people should buy and dispose of less stuff. Michael Bloch is a consultant and blogger. He writes, "There is concern building that

Recycling Adds to Overall Emissions

Writer Andrew Handley argues that the environmental benefits of recycling have been exaggerated by recycling advocates. According to Handley, air pollution is just one of the negative effects of recycling:

The recycling process itself produces a lot of pollutants—from the exhaust billowing out of recycling trucks to energy used at recycling plants. In 2009 there were about 179,000 waste collection vehicles on the road—that's both recycling and garbage collection. The exhaust from each one of those vehicles contains over three dozen airborne toxins.

The thing is, you can't separate garbage trucks from recycling trucks—there's no lesser evil. They both run on fossil fuels, and they both produce exhaust. By adding more trucks to the fleet, no matter what their purpose, we're increasing air pollution.

Andrew Handley, "Ten Ways Recycling Hurts the Environment," Listverse.com, January 27, 2013.

the recycling wave is allowing us to still be rampant consumers, a throwaway society; and recycling [is] some justification for maintaining this mindset."[8] Bloch thinks that buying fewer goods or reusing more goods may be a good way to reduce waste. He says the best way is to use fewer goods in the first place.

Amy Perlmutter is a national expert in waste management. She says, "You can't give people a recycling bin and expect a light bulb to go on."[9] She says the best way to boost recycling may be to think about the future. Randi Mail agrees. She is a recycling director. "We need to change consumption habits, change products, and shift public behavior away from disposable items," she says. "That doesn't necessarily mean sacrifice, it just means change."[10]

Closing Arguments

The recycling debate continues. People do not agree on the environmental impact of recycling. Advocates think that recycling is a big part of protecting the planet. Others feel differently. They think that the benefits of recycling have been exaggerated.

Examine the Opinions

Using Facts to Support an Argument

A main idea found in this chapter is how much recycling helps the planet. Both supporters and critics of recycling say that some forms of recycling have benefits. Some people think that recycling has a big impact on lowering greenhouse gases and pollution. They also say it conserves natural resources and energy. Others say that these benefits are offset by the emissions created and energy used by recycling.

Using facts is one technique to support an argument. When examining this issue, both sides of the argument use facts to support their arguments. On one side, recycling advocates say that recycling saves resources. The factual statements "Recycling paper saves trees. It provides 37 percent of the raw materials used to make new paper products" support this position. On the other side, critics use another fact to support their point of view: "Most of the pulp from paper comes from sources that are grown for paper. If the demand for paper dropped, the businesses that own these trees would find another way to use them." Both statements are factual. But the way each side uses the information proves its point. This is a good example of using facts to support an argument.

3 Is Recycling Economically Efficient?

👍 **Yes:** Recycling Is Efficient

Many towns and cities are struggling with their budgets. Some people think that recycling programs cost too much. They say that these programs are just too costly and inefficient to be successful. Some think that the money would be better spent in other areas, such as schools or police forces. Others think that these programs can create economic benefits for cities and towns. They say that these benefits outweigh the program costs.

Creating Jobs and Revenue

Some people say that recycling is a big part of the economy. They say that it gives a lot of people jobs. These jobs range from materials handling and processing to product manufacturing. In 2013 the Institute of Scrap

Workers at a recycling plant. Recycling creates jobs.

Recycling Industries (ISRI) did a study. It found recycling creates nearly half a million jobs in the United States. It spurs more than $87 billion a year in financial activity. Robin Wiener works for the ISRI. He says: "The scrap recycling industry serves as an economic driver for our country, a job creator, and major exporter. At a time when the national job market is fragile, the recycling industry is creating high-quality, good-paying opportunities for Americans. This in turn feeds money back into the economy."[11] The recycling industry also creates jobs and revenue for companies that buy used goods, recycle them, and sell them as new products.

Recycling saves money. It lowers waste disposal costs. And businesses can sell their recyclables for money. In 2013 the New Jersey WasteWise Business Network put out a report. It said recycling saves thousands of dollars in that state. In 2011 one company in the state saved more than $145,000. It did this by recycling items instead of throwing them out. Also in 2011, another plant in New Jersey saved more than $85,000 through its recycling programs.

A New Jersey plant recycles plastics into new plastic bottles.

The Global Economy of Recycling

Trash has increased around the world. And more businesses have sprung up to deal with recycling. Adam Minter is the author of *Junkyard Planet*. He says that recycling has a big impact on the world's economy. "As Asia's demand for raw materials grew, at the same time the amount of stuff being thrown out in developed countries like the U.S. and the EU and Japan grew, and as that stuff grew, the industry that was devoted to collecting it, recycling it and, to a large extent, exporting it grew with it," he says. "And so the employment opportunities and the wealth generated in the developed world and in the developing world has been extraordinary."

Quoted in Fresh Air, "Christmas Lights Make Slippers in Global 'Junkyard' Economy," National Public Radio, November 13, 2013.

Product Life Cycle Counts

Some people say it costs less to dump waste into a landfill. Others say that idea is misleading. They say the true cost of waste disposal must be considered. The cost of making a product from recycled items should be compared with the cost of making the same thing from new materials. Then recycling becomes **cost-effective**. Edward Humes is an author. He writes:

> Recycling economics should be weighed not as the last step in waste disposal, but as the first step in manufacturing. The proper comparison is not landfill vs. recycling, but virgin materials vs. recycled ones. Which is more costly to manufacturers of products? The answer generally favors recycled materials, as there is enormous demand for recycled metals, paper, and certain types of plastic. Here, both the environmental and economic benefits of recycling are revealed: there are substantial energy and carbon savings that come with the choice of many recycled raw materials over virgin ones. Recycling aluminum cans, for instance, saves a whopping 96 percent of the energy needed to produce aluminum from bauxite ore.[12]

But Not So Fast...

 No: Recycling Costs Too Much

Some people think that recycling programs are just too costly. They say that recycling costs millions of

A worker prepares alumnium cans for recycling.

dollars. Cities pay to pick up, sort, and process items. The volume of these items needs to be great enough to pay for the separate trucks, centers, and processing. That way recycling will make financial sense.

In 2010 Ocean City, Maryland, dropped its recycling program. The city said it was too costly. Richard Malone works at the city's Public Works Department. He says it cost the city $394 to haul off a ton of recyclables. But a ton of trash cost only $162 to haul off. The decision to drop the program will save the city around $1 million a

Community recycling programs can be costly.

year. Jennifer Berry works for Earth911. She says that other towns are also looking at the cost of their recycling programs. "People like to think of recycling as a feel-good industry, but these are commodities and people are having a harder time selling them," she says. "A million dollars is a lot of money in any town right now."[13] When communities do not recycle, Berry says that more people will have to take responsibility for recycling themselves.

Complex and Costly

Some materials, such as plastics, are too hard to recycle efficiently. There are many types of plastics. They are used in a wide range of products from bottles to yogurt cups. Plastic items are labeled with a number from 1 to 7. The number depends on the **polymer** of the plastic. Each type can only be recycled with its own

Each type of plastic can only be recycled with its own kind, which adds to the overall costs.

kind. Sometimes, even two items with the same number cannot be recycled together. They may melt at different temperatures. So plastics must be sorted carefully, sometimes by hand, before they can be recycled. This adds to the cost of recycling plastic.

Even when sorted, many plastic items have additives. These make it impossible to recycle the plastic to its original state. These additives can ruin a batch of

No Value

For some items, recycling loses money. San Francisco's Department of Waste says it costs $4,000 to recycle 2,205 pounds (1,000kg) of plastic bags. The resale price of the recycled product was $32. Joseph Gho works at a company that makes biodegradable plastics. He says that recycling does not always save money. "Nobody wants it. There's no value. It doesn't make sense," he says. "Besides the financial, the economic cost, you've got the environmental cost" of recycling unwanted material. "The trucks running out there, burning fuel ... you have to use energy, you've got CO2 emissions."

Quoted in Kevin Libin, "The Recycling Conundrum: How Your Blue Bin Hurts the Environment," *National Post*, December 4, 2009.

plastics waiting to be recycled. If ruined, the entire batch is thrown out. Therefore, some think it may cost less to dump these items into a landfill or burn them in an incinerator.

Time Is Money

Recycling tends to takes more time and effort than just throwing out trash. Darren works for a nonprofit group in Washington, D.C. He says that people should not forget the cost of time spent when they evaluate the cost of recycling.

> *What I wish everyone would learn in Economics 101 is that there are trade-offs in life. There are both benefits and downsides to recycling," he says. "Individually, time is the most precious resource we use when we recycle. You could have done something else with that time used to recycle, and you can never get back spent time. On the city level, it's time, effort and money. It is a question of whether recycling is the best use of that money, or if it would be better spent on education or health care. There are always trade-offs.[14]*

Closing Arguments

The debate over the costs and benefits of recycling will likely continue. New recycling technologies will affect the costs of recycling programs. The rise and fall of raw material costs will affect the cost of such programs. So will the changing costs of waste disposal. These costs are different all over the country. So cities and towns must make their own decisions about the economics of recycling.

Examine the Opinions

Testimonials

In this chapter the author quotes waste professionals. They give their opinion about the cost of recycling. One of these is Robin Wiener. He thinks that recycling is good for the United States. He says that recycling creates good jobs for Americans. When a writer quotes an expert, it is a technique called a **testimonial**. A testimonial offers further evidence or proof of an opinion. When evaluating a testimonial, it is important to consider the source. Is the source an expert in the subject? Is he or she part of a group that is an authority on the subject? On the other hand, is the source biased about the topic? Does he or she have something to gain by supporting one side of an issue? Evaluating the source will help you decide how reliable the testimonial is. It is important to keep in mind that an expert's opinion can be biased. For example, a person who works for an environmentalist group such as Greenpeace will hold certain views about pollution. A person affected by a certain disease may be biased about health care policy. This does not make his or her opinion unworthy of consideration. But it is important to know the person's bias and take that into account.

4 Should Recycling Be Mandatory?

Yes: Recycling Should Be Mandatory

In 2005 Seattle passed a mandatory recycling law. The law bans recyclables in the trash of homes and businesses. Homes must sort and recycle basic items. This means paper, cardboard, aluminum, glass, and plastic. Businesses must recycle all paper, cardboard, and yard waste.

Homes and businesses must follow this law. If they do not, they may be fined. Brett Stav works for Seattle Public Utilities. He explains the penalties for not obeying the law:

> For businesses, if we find more than 10 percent of the garbage container is filled with things like paper or cardboard, we'll leave a tag. On the third tag, we'll leave a $50 fine. On apartments, it works the same way. For households, we don't

A worker breaks down electronics for recycling at a Seattle plant. In 2005 Seattle passed a mandatory recycling law.

fine anyone. Just automatically if we find too many recyclables in your garbage, we'll leave a tag and ask you to sort it out and then we'll collect the garbage can the next week.[15]

Stav says that most people follow the new law. "When we first started out, we had more than 90 percent of apartments and businesses complying with the new ordinance. The majority of residents get recycling and hardly any garbage cans were left behind at all. We've seen that trend continue."[16] In addition, Seattle

A garbage hauler leaves a tag on a garbage bin because it contains recyclables.

has considered additional legislation that would make it mandatory for businesses to recycle glass, plastic, tin, and aluminum.

Mandatory Recycling Successes

Mandatory recycling has worked in many cities. In Seattle the recycling rate was 38 percent in 2003. After mandatory recycling, the rate rose to 55.7 percent in 2013. Cleveland started a mandatory program in 2011. "This program benefits all involved," says Ronnie Owens. He works at the Cleveland Division of Waste. "The city gains revenue from recyclables, which helps the bottom line. Workers experience fewer

National Mandatory Recycling

Today the United States does not have a national recycling program. Ricardo Llerandi-Cruz is a member of the Puerto Rico House of Representatives. He thinks that it is time for the United States to pass recycling laws. "As a nation, we need to set forward a strategic vision for this problem," he writes. "The most important decision we can make is the establishment of a national mandate. Recyclable material would be prohibited from all garbage centers…. Mandatory recycling is a hard sell in the United States, where our economy runs largely along free market lines. But a national mandate is more than important, it's imperative, if we are going to deal with the matter of waste disposal seriously. It's time Congress recognizes this."

Ricardo Llerandi-Cruz, "Time for a National Recycling Mandate," *The Hill* (blog), November 2, 2013.

injuries because of the automated process. Citizens receive new, wheeled carts at no cost. And, Cleveland raises its profile as an environmentally friendly city."[17]

Some mandatory programs give a fine to those who do not follow the rules. In Cleveland the city keeps track of people's recycling with radio frequency identification.

San Francisco law requires separate bins for composting, recycling, and trash.

They attach tags to the recycling carts. "We know each time we tip the recycling cart," Owens says. "If they haven't recycled in four weeks time, we'll review those records and the next time we go out to their home and they still haven't recycled, they'll be issued a violation. That violation is basically a $100 fine."[18]

In 2009 San Francisco passed a new composting and recycling law. It is the strictest in the country. The law says every home and business must have separate color-coded bins for recycling, compost, and trash. If waste is not sorted right, a warning will be issued. After a few warnings, the city can charge a fine.

Many people think that mandatory recycling helps cities. They say that these programs boost recycling rates. This helps keep the planet clean. It also reduces landfill disposals. The programs save money on trash disposal. And they create local jobs. Travis Schulz is chair of the Missouri Valley Resource Council. He says, "Governments are here to help protect all people and encourage the right behavior. Mandatory recycling saves all of us money in the long run, creates a healthier society and more conscious and responsible citizens. It is the right thing to do."[19]

Voluntary Programs Do Not Work

Many people think that recycling should be mandatory. They say voluntary programs do not work. Schulz says that such programs do little to reduce the volume of trash. "Allowing an 'opt-out' option really defeats the main purpose of curbside recycling, which is to reduce the amount of garbage in our landfill," he says. "The problem isn't with those who are already recycling, but with those who aren't. Opting out allows the polluters to not only continue to fill up our landfill unnecessarily, but also to be rewarded financially for this bad practice."[20]

But Not So Fast...

 No: Recycling Should Not Be Mandatory

Mandatory recycling may boost recycling rates. But some think that recycling should be a personal choice. It should not be forced by the government. They claim that mandatory recycling is a violation of people's liberty. In San Francisco the thought of inspectors going through trash bins upsets some people. Sean Elsbernd is a worker for the city. He says, "This is a little too much big brother, even for me. We've got a huge problem in my district and a lot of other parts of the city with people who go in and out of garbage cans at night scavenging. ... Are we creating a whole brand-new problem?"[21]

A recycling can on a city street encourages pedestrians to recycle.

Matt is a college graduate who lives in Utah. He chooses not to recycle. He does not agree with mandatory recycling. "I'm sure that they pay for themselves to some degree, but I am annoyed that my tax dollars go to recycling programs," he says. "If people are into recycling, they should do it on their own. It's not the government's place to decide which causes I support."[22]

Recycle Resources, Not Trash

Some people point out that mandatory recycling does not work well. When it is voluntary, people recycle items that have value. This saves useful resources. But mandatory programs cause people to recycle everything. They even recycle things that are not useful. Michael C. Munger is an economics professor.

Recycling Is an Individual Choice

Some say that the government should not make recycling decisions. Instead, people themselves should decide what is right for them. Craig Kohtz is a writer. He says that recycling should be voluntary. "The reason we pay for curbside recycling isn't because it makes economic sense or even environmental sense. It's because it is convenient and it makes us feel better about the environment," he says. "Many of us are willing to pay for this feeling, which is a legitimate reason to recycle. However, forcing others to pay more for the privilege of recycling so we can feel better about ourselves is not."

Craig Kohtz, "Why We Shouldn't Force Recycling," *Lincoln (NE) Journal Sta*r, August 9, 2013.

He says, "If we actually cared about resources and the environment, we would be better off immediately eliminating all mandatory recycling programs nationwide. We waste more energy, and cause more pollution, by trying to recycle than we would if we abandoned these programs and let market forces pull what is valuable from the waste stream."[23]

Pay-as-you-throw programs could encourage people to sort their recyclables, compost, and trash in order to pay less for trash hauling.

A Better Way to Encourage Recycling

Some think that pay-as-you-throw trash programs are a good way to boost recycling. They work better than regular trash services that charge homes and businesses a flat fee for collection, hauling, and dumping of trash. With pay-as-you-throw, people are charged by the amount of trash they throw out. Those with several large trash cans are charged more than those with one small can. People pay for what they use. And these systems do not charge for picking up recyclables.

Many people support programs like these. They say that more people will recycle if it is voluntary. These programs make it cost less to recycle than to throw out trash. "These communities provide consumers with an economic incentive to waste less and recycle more and, as a group, they recycle 30 to 40 percent more than the rest of the country,"[24] says Edward Humes.

Closing Arguments

In order to boost recycling rates, some cities have passed mandatory recycling laws. They think that these laws are the best way to get people to recycle. Others do not agree. They think that the choice to recycle should be voluntary. They say programs that help people reduce the amount of trash they create are better ways to reduce waste.

Examine the Opinions

Understanding the Difference Between Fact and Opinion

This chapter presents information and viewpoints from many people on mandatory recycling. It is important to know the difference between a fact and an opinion. A fact cannot be argued. An opinion is one person's view on an issue. Matt said, "If people are into recycling, they should do it on their own. It's not the government's place to decide which causes I support." This is an opinion. The chapter also quotes Edward Humes. He said, "These communities provide consumers with an economic incentive to waste less and recycle more and, as a group, they recycle 30 to 40 percent more than the rest of the country." This statement presents a fact. When looking at an argument, facts may carry more weight than opinions. Facts are not always good. And opinions are not always bad. But facts are easier to prove. So using facts helps make an argument stronger.

Wrap It Up!

Write Your Own Essay

In this book, the author gave many opinions about recycling. These opinions can be used to write a short essay on recycling. Short opinion essays are a common writing form. They are also a good way to use the ideas in this book. The author gave several common writing techniques and evidence that can be used. Using facts to support an argument, testimonials, and understanding the difference between fact and opinion were techniques used in the essays to sway the reader. Any of these could be used in a piece of writing.

There are 6 steps to follow when writing an essay:

Step One: Choose a Topic

Choose a topic to write about in your essay. You can start with one of the three chapter questions presented in this book.

Step Two: Research Your Topic

Decide which side of the issue you will take. After choosing your topic, use the materials in this book to write the thesis, or theme, of your essay. You can use the articles and books cited in the notes and also the bibliography. You could also interview people in your life who recycle or do not recycle and quote them in your essay.

Step Three: Write Your Theme

The first paragraph should state your theme. For example, in an essay titled "Recycling Should Be Mandatory," you would state your opinion. Say what action you think should be taken to make recycling mandatory and why. You could also use a short anecdote, or story, that proves your point and will interest your reader.

Step Four: The Body of the Essay

In the next three paragraphs, develop this theme. You should come up with three reasons why recycling should be mandatory. For example, three reasons could be:

- *Recycling reduces trash thrown out in landfills.*
- *Recycling creates jobs for people.*
- *Recycling saves natural resources.*

These three ideas should each be given their own paragraph. Be sure to give a piece of evidence in each paragraph. This could be a testimonial from an environmental scientist or waste program expert. It could be a fact from a study, poll, or government group. Each paragraph should end with a transition sentence that sums up the main idea in the paragraph and moves the reader to the next.

Step Five: Write the Conclusion

The final, or fifth, paragraph should state your conclusion. The conclusion should restate the theme and sum up the ideas in your essay. It could also end with a good quote or piece of evidence that wraps up your essay.

Step Six: Review Your Work

Finally, be sure to reread your essay. Does it have quotes, facts, and/or anecdotes to support the conclusions? Are the ideas clearly presented? Have another reader take a look at your project in order to see whether he or she can understand your ideas. Make any changes that you think can help make your essay better.

Congratulations on using the ideas in this book to write a personal essay!

Notes

Chapter 1: Why Is Recycling an Issue?

1. Quoted in Larry McShane, "Recycling May Halt in New York," BG News (Bowling Green State University), April 23, 2002. www.bgnews.com/recycling-may-halt-in-new-york/article_b1159eb8-de5c-5d95-91ea-8d43eec36032.html.
2. Quoted in McShane, "Recycling May Halt in New York."

Chapter 2: Is Recycling Good for the Environment?

3. Quoted in Plastics Make It Possible, "Spotlight on Patty Moore, Recycling Expert," November 2010. http://plasticsmakeitpossible.com/2010/11/spotlight-on-patty-moore-recycling-expert.
4. Quoted in State of New Jersey Department of Environmental Protection, "Christie Administration Honors New Jersey's Recycling Leaders," October 24, 2013. www.nj.gov/dep/newsrel/2013/13_0097.htm.
5. Quoted in Rainbow Light, "Recycling—the Carbon Footprint on Food and Packaging." www.rainbowlight.com/about-us-recent-news-recycling-the-carbon-footprint-on-food-and-packaging.aspx.
6. Drew Thornley, "Energy and the Environment: Myths and Facts," Manhattan Institute, April 2009. www.manhattan-institute.org/energymyths/myth5.htm.
7. Todd Myers, "Eco-fads: Does Recycling Save Trees? Seattle City Light Gets It Wrong," Washington Policy Center, April 4, 2012. www.washingtonpolicy.org/blog/post/eco-fads-does-recycling-save-trees-seattle-city-light-gets-it-wrong.

8. Michael Bloch, "Think Reuse Before Recycle," Green Living Tips, October 6, 2010. www.greenlivingtips.com/articles/reuse-vs-recycle.html.

9. Quoted in Barbara Moran, "Are Big Blue Bins Bad for Recycling?," *Boston Globe,* July 14, 2013. www.bostonglobe.com/magazine/2013/07/13/does-recycling-really-work/qo9I5UM6yXw4ouswmIdXFK/story.html.

10. Quoted in Moran, "Are Big Blue Bins Bad for Recycling?"

Chapter 3: Is Recycling Economically Efficient?

11. Quoted in Institute of Scrap Recycling Industries, "New Economic Impact Study Shows Recycling Industry as a 'Thriving Economic Engine,'" July 16, 2013. www.isri.org/ISRI/Whats_New/2012/New_Economic_Impact_Study_Shows_Recycling_Industry.aspx.

12. Edward Humes, "Recycling: Why Better than Nothing Isn't Good Enough," Cato Unbound, June 7, 2013. www.cato-unbound.org/2013/06/07/edward-humes/recycling-why-better-nothing-isnt-good-enough.

13. Quoted in Meredith Cohn, "To Save Money, Ocean City Drops Recycling Program," *Baltimore Sun*, April 21, 2010. http://articles.baltimoresun.com/2010-04-21/features/bs-gr-ocean-city-recycle-20100420_1_recycling-program-drop-off-ocean-city.

14. Quoted in Ashley Schiller, "Why People Don't Recycle," Earth911, October 25, 2010. http://earth911.com/news/2010/10/25/why-people-dont-recycle.

Chapter 4: Should Recycling Be Mandatory?

15. Quoted in Jennifer Langston, "Mandatory Recycling Program Working Well," *Seattle Post-Intelligencer*, March 14, 2006. www.seattlepi.com/local/article/Mandatory-recycling-program-working-well-1198413.php.

16. Quoted in Langston, "Mandatory Recycling Program Working Well."

17. Quoted in Ken Prendergast, "Recycling Program Began This Week in Cleveland," Cleveland.com, September 8, 2011. www.cleveland.com/sunpostherald/index.ssf/2011/09/recycling_program_began_this_w.html.

18. Quoted in Vince Bond, "Mandatory Recycling Laws Grow in Popularity," Waste Recycling News, May 30, 2012. http://vincebond.wordpress.com/category/waste-recycling-news.

19. Travis Schulz, "Recycling Should Be Mandatory," *Bismark (ND) Tribune*, February 24, 2013. http://bismarcktribune.com/news/columnists/recycling-should-be-mandatory/article_533bf2d0-7d06-11e2-975e-0019bb2963f4.html.

20. Schulz, "Recycling Should Be Mandatory."

21. Quoted in John Cote, "SF OKs Toughest Recycling Law in US," *San Francisco Chronicle*, June 10, 2009. www.sfgate.com/green/article/S-F-OKs-toughest-recycling-law-in-U-S-3295664.php.

22. Quoted in Schiller, "Why People Don't Recycle."

23. Michael C. Munger, "Bootleggers, Baptists, and Recyclers," Cato Unbound, June 13, 2013. www.cato-unbound.org/2013/06/13/michael-c-munger/bootleggers-baptists-recyclers.

24. Humes, "Recycling."

Glossary

additives: Substances that are added to another substance to change it.

composting: Making a mixture of rotted leaves, vegetables, manure, and other organic materials to create fertilizer.

cost-effective: Describes something that produces good results for the amount that it costs.

decompose: To rot or decay.

greenhouse gas: Gases that trap heat in the Earth's atmosphere and have been linked to global warming.

incineration: The burning of trash or waste materials.

mandatory: Ordered or required.

organic: Having to do with or coming from living things.

polymer: A natural or synthetic compound made up of small, simple molecules linked together in long chains of repeating units.

recycling: Processing old items such as glass, plastic, newspapers, and aluminum cans so that they can be used to make new products.

testimonial: A statement in support of a particular truth, fact, or claim.

textiles: Fabric or cloth that has been woven or knitted.

Bibliography

Books

Edward Close, *What Do We Do with Trash?* New York: PowerKids, 2013. This book talks about types of trash and ways to recycle.

Rebecca Hunter, *Waste and Recycling.* Mankato, MN: Sea-to-Sea Publications, 2012. This book explains the issues surrounding waste and recycling and gives information about the ways materials can be recycled and how recycling can help the environment.

Nick Winnick, *Reduce Waste: Being Green.* New York: AV2 by Weigl, 2011. The author presents information about being green and ways to reduce waste.

Articles

Meera Dolasia, "Will 'Recycled Island' Finally Become a Reality?," DOGO News, October 6, 2013. www.dogonews.com/2013/10/6/will-recycled-island-finally-become-a-reality. This article talks about the idea of Dutch architect Ramon Knoester to recycle plastic that is polluting the ocean and turn it into an island where people could live.

Catherine Clarke Fox, "Drinking Water: Bottled or from the Tap?," National Geographic Kids. http://kids.nationalgeographic.com/kids/stories/spacescience/water-bottle-pollution. This article discusses the issue of bottled water versus tap water and how using each impacts recycling and the environment.

Kochava R. Greene, "What Happens to Trash After It Is Hauled Away?," eHow. www.ehow.com/how-does_5187578_happens-trash-after-hauled-away_.html. This article explains what happens to trash after it is hauled away by sanitation trucks: It is taken to an incinerator, a landfill, or a recycling center.

Websites

National Institute of Environmental Health Science Kids' Pages (http://kids.niehs.nih.gov/exp). This website has a "Reduce, Reuse, and Recycle" section that gives young readers information about recycling. It also has links to more recycling information, games, and quotes on the environment and nature.

Environmental Protection Agency Recycle City (www.epa.gov/recyclecity). Kids can explore Recycle City to see how residents reduce waste, use less energy, and save money by reducing, reusing, and recycling.

Recyclezone (www.recyclezone.org.uk). This website is sponsored by the National Institute of Environmental Health Sciences in the United Kingdom. It provides great information that is presented in an easy-to-read format. The site is interactive, with many great jokes, brain teasers, music, and games.

Recycling Facts (http://recyclingfacts.org). This website has many facts about recycling, as well as helpful information about ways to reduce, reuse, and recycle.

Index

About the Author

Carla Mooney is an author of many books for young readers. She loves learning about issues and understanding different opinions. A graduate of the University of Pennsylvania, she lives in Pittsburgh, Pennsylvania, with her husband and three children.